Let's Investigate!
Hands-on Science

To the Teacher

Help students explore the wonders of science with the mind-stretching activities in this exciting hands-on science series. Each book includes:

- Fun, easy-to-prepare activities that cover topics from each of the three main branches of science: Physical Science, Earth Science, and Life Science
- Clear, step-by-step instructions that foster both group and independent learning
- Guided questions for developing observation and critical thinking skills
- Fascinating facts and extension activities that further challenge students' thinking and enrich their learning

Background information and teaching tips are also included in the Activity Guide at the back of the book. This information will help you guide your students through each activity and ensure their understanding of the scientific concepts.

How to Use the Activities

The activities in this book are perfect for both group learning and independent study. Most require simple, readily available materials and can be easily adapted to your needs.

While a few of the activities are only one page in length, most consist of two pages—an instruction sheet and a record sheet, where students can record their data and observations. The following are ideas for ways to use this book:

Whole-Class/Individuals—Use the activities for whole-class learning. Reproduce copies of both pages of an activity for each student, and let students carry out the activities independently.

Whole-Class/Pairs or Groups—Let the class work in pairs or small groups. Provide record sheets for every student, but reproduce only one instruction sheet for each pair or group.

Learning Centers—Reproduce one instruction sheet and laminate it. Display the sheet at a learning center. Reproduce record sheets for every student, and leave them at the center as well. Let one group at a time work at the center. After all students have completed an activity, discuss the results with the class.

Demonstrations—Conduct the activities as class demonstrations. Pass out record sheets to your students, and have them write their observations and discoveries.

Name _____   Magnets

Magical Attraction

Let's Find Out

How far does a magnet's power reach?

What You'll Need

- wand magnet
  (or long bar magnet)
- paper clip
- shoebox
- thread
- scissors
- tape
- ruler

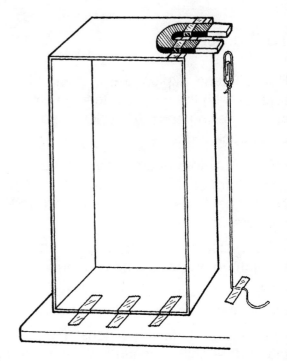

What to Do

Work with a partner.
1. Tape the shoebox upright on a table.
2. Tape the magnet on top of the box, with one inch of it extending over the edge.
3. Cut a piece of thread a little longer than the height of the box. Tie one end to the paper clip.
4. Touch the paper clip to the bottom of the magnet. Gently pull down on the thread so that there is a bit of space between the magnet and the paper clip. As you hold the thread, have your partner tape it in place.
5. Carefully tug on the thread to lower the paper clip. Measure how far away it can go from the paper clip before it drops.

What You Saw

How far away was the paper clip from the magnet before it fell? Write your answers on your record sheet.

Think About It

Two forces were acting on the paper clip. One was the force of the magnet. What was the other force? Why did the paper clip eventually fall? Write your answers on your record sheet.

© Milliken Publishing Company

Name _____

Magnets

Magical Attraction

Let's Find Out

How far does a magnet's power reach?

What You Saw

How far away could you move the paper clip from the magnet before it fell?

_____

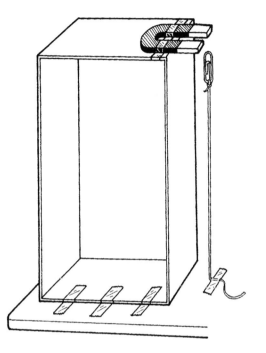

Think About It

Two forces were acting on the paper clip. One was the force of the magnet. What was the other force?

_____
_____

Why did the paper clip eventually fall?

_____
_____
_____

Try This!
Set up your experiment again. See if the paper clip will still "float" if you put a piece of paper between it and the magnet. Try other materials besides paper, too!

© Milliken Publishing Company   3   MP3501

Name _____   Static electricity

Together or Apart?

Let's Find Out

What happens when you rub two balloons with wool and place them near each other?

What You'll Need

- two balloons
- wool cloth
- string
- tape

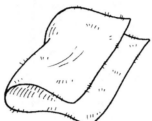
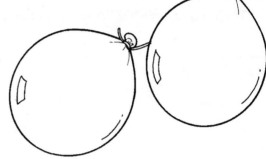

What to Do

1. Blow up two balloons. Tie a 12-inch length of string to each one.
2. Tape the strings to the edge of a table so that the balloons hang down two inches apart.
3. Rub one balloon with the wool cloth. Rub quickly 30 times. Carefully put the balloon back so that it does not sway a lot while it hangs. Repeat the activity with the second balloon.
4. Let both balloons hang, and see what happens.
5. Hold the string of one balloon. Move that balloon close to the other one without the two touching. See what happens to the second balloon.

What You Saw

How did the two hanging balloons behave after they were rubbed with wool? What happened when you moved one balloon toward the other one? Write your answers on your record sheet.

Think About It

When you rubbed the balloons with the wool, they got an electric charge. Do you think objects with the same kind of electric charge attract (draw near to) each other or repel (push away) each other? Write your answer on your record sheet.

© Milliken Publishing Company

MP3501

Name _____   Static electricity

Together or Apart?

Let's Find Out

What happens when you rub two balloons with wool and place them near each other?

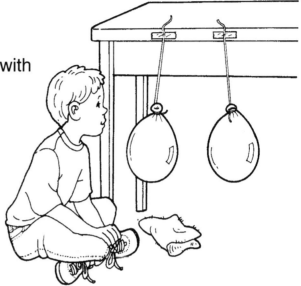

What You Saw

How did the two hanging balloons behave after they were rubbed with wool?

_____

_____

What happened when you moved one balloon toward the other one?

_____

Think About It

When you rubbed the balloons with the wool, they got an electric charge. Do you think objects with the same kind of electric charge attract (draw near to) each other or repel (push away) each other?

_____

Try This!
Spread 10 half-inch paper squares on a table. Rub a balloon with a wool cloth. Rub quickly about 30 times. Then pass the balloon over the squares without touching them. What happens to the squares?

Name _____   Static electricity

Jumping Race

Let's Find Out

How does an object that is charged with static electricity affect grains of salt and pepper?

What You'll Need

- small plate
- salt shaker
- pepper shaker
- balloon
- wool cloth

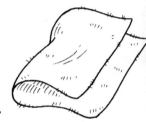

What to Do

1. Sprinkle a little salt and pepper onto the plate.
2. Blow up the balloon and tie off its end.
3. Hold the balloon close to the plate. See if anything happens.
4. Rub the wool over the balloon to charge it with static electricity. Rub 30 times very fast.
5. Hold the balloon over the plate. Slowly bring it downward, but don't let it get too close to the salt and pepper. Check the balloon's surface.
6. Bring the balloon closer toward the plate. Check the balloon's surface again.

What You Saw

What happened when you first held the balloon over the plate?
What happened when you held the charged balloon over the plate?
Which jumped up first—the salt or the pepper? Write your answers on your record sheet.

Think About It

What was attracted to the charged balloon—the salt, the pepper, or both?
Which is lighter—grains of salt or grains of pepper? How do you know?
Write your answers on your record sheet.

© Milliken Publishing Company   6   MP3501

Name _____

Static electricity

Jumping Race

Let's Find Out

How does an object that is charged with static electricity affect grains of salt and pepper?

What You Saw

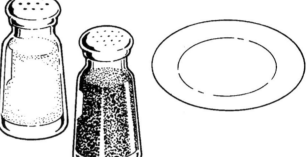

What happened when you first held the balloon over the plate?

_____

_____

What happened when you held the charged balloon over the plate?

_____

Which jumped up first—the salt or the pepper? _____

Think About It

What was attracted to the charged balloon—the salt, the pepper, or both?

_____

Which is lighter—grains of salt or grains of pepper? How do you know?

_____

_____

© Milliken Publishing Company

MP3501

Name _____

Electricity

Make a Circuit

Let's Find Out

How can you make an electric circuit with a light bulb, a battery, and some wire?

What You'll Need

- C or D battery
- flashlight bulb
- tape
- 12-inch piece of flexible, insulated wire (1/4 inch stripped at one end and 2 inches stripped at the other end)

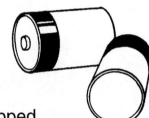

What to Do

1. Pick up the end of the wire that has 1/4 inch of it stripped. Tape it to the bumpy end (the positive end) of the battery.
2. Coil the other end of the wire around the base of the bulb.
3. Touch the bottom of the bulb to the flat end (the negative end) of the battery.

What You Saw

What happened to the bulb when you touched it to the battery?

_____

Think About It

You just made an electric circuit—a path through which electric charges can flow. Look at your circuit. What is the source of electric charges?

Try This! Make your circuit as before. Then add a paper clip. Use one hand to hold the paper clip to the flat end of the battery. Use your other hand to bring the bulb to the paper clip. Does the bulb light up?

© Milliken Publishing Company

Name _____

Electricity

Cutlery Circuit

Let's Find Out

Can a fork and spoon be used to make an electric circuit?

What You'll Need

- C or D battery
- flashlight bulb
- stainless steel fork
- stainless steel spoon
- tape

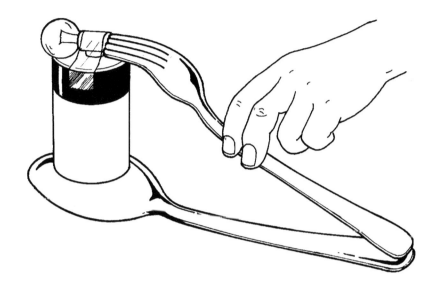

What to Do

1. Tape the bulb sideways to one end of the battery. Make sure the bulb is visible.
2. Place the back of the spoon on the other end of the battery.
3. Place the handle of the fork on the handle of the spoon.
4. Touch the bulb's base with the prongs of the fork.

What You Saw

What happened when you touched the bulb with the fork?

_____

Think About It

Materials that let electricity pass through easily are called conductors. Materials that do not let electricity pass through easily are called insulators. Are the fork and spoon in your experiment conductors or insulators?

_____

© Milliken Publishing Company   MP3501

Name _____   Liquids—density

Ice Cube Drops

Let's Find Out

What happens when you put an ice cube in cooking oil?

What You'll Need

- small, clear cup (or jar)
- cooking oil
- ice cube

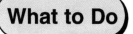

What to Do

1. Fill the cup three-quarters full with oil.
2. Put an ice cube in the cup. See what happens to the ice cube.
3. Check the cup after five minutes. Notice what is happening to the ice cube.
4. Check the cup after half an hour and after an hour.

What You Saw

What happened to the ice cube when you first put it in the oil?
What happened afterward?
Write your observations on your record sheet.

Think About It

Is an ice cube heavier or lighter than oil?
Is water heavier or lighter than oil?
Write your answers on your record sheet.

© Milliken Publishing Company   MP3501

Name _____   Liquids—density

Ice Cube Drops

Let's Find Out

What happens when you put an ice cube in cooking oil?

What You Saw

What happened to the ice cube when you first put it in the oil?

_____

Describe what the ice cube looked like each time you checked the cup.

After 5 minutes:

_____

After half an hour:

_____

After an hour:

_____

Think About It

Is an ice cube heavier or lighter than oil?

_____

Is water heavier or lighter than oil?

_____

Try This! Make a "liquid sandwich." Pour some molasses or syrup into a baby food jar. Add some colored water a spoonful at a time by dribbling it down the sides of the jar. Then add spoonfuls of cooking oil.

© Milliken Publishing Company   11   MP3501

Name _____   Solids—density

A Fruity Swim

Let's Find Out

Will an orange sink or float in a bowl of water?

What You'll Need

- medium-sized bowl
- orange
- paper towels

What to Do

1. Fill the bowl with water.
2. Put an orange in the bowl. See if it floats or sinks.
3. Take the orange out of the bowl and peel it. Use paper towels to wipe up any mess.
4. Put the orange back in the bowl. See if it floats or sinks.

What You Saw

What happened when you put the unpeeled orange in the bowl? _____

What happened when you put the peeled orange in the bowl? _____

Think About It

Why did the orange act differently each time?

_____

_____

Did You Know?
Oranges probably first grew in Asia. Traders then took oranges to other parts of the world. In 1493, Christopher Columbus was the first to bring oranges to America.

© Milliken Publishing Company   MP3501

Name _____   Balance

Impossible Moves

Let's Find Out

Can you get "stuck" when you try to lift your foot or pick up an object?

What You'll Need

- ball
- wall

What to Do

Activity 1
1. Stand next to a wall so that your right shoulder touches the wall.
2. Move your left foot a few inches to the left. Put your right foot against the wall.
3. Lift your left foot.

Activity 2
1. Put a ball on the floor, several inches away from the wall.
2. Stand with your heels touching the wall
3. Pick up the ball without moving your feet.

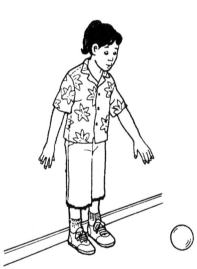

What You Saw

What happened when you tried to lift your foot or pick up the ball?

_____

Think About It

Why do you think you got "stuck" when you tried to do the activities?

_____

Name _____

Balance

A Balancing Act

Let's Find Out

How can you keep a ruler balanced as it rests on a point?

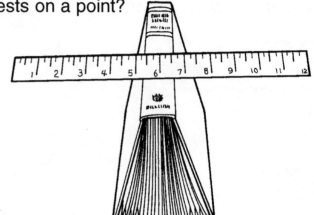

What You'll Need

- book with a 1/2" spine
- 12-inch ruler
- 3 pennies

What to Do

1. Open the book and set it on a table so that its spine faces up.
2. Lay the ruler across the book's spine and balance it. Notice what part of the ruler rests on the spine.
3. Place a penny on one end of the ruler. Move the ruler so that it stays balanced.
4. Place a second penny on the opposite end of the ruler. Move the ruler to keep it balanced.
5. Place two pennies on one end of the ruler and one penny on the other. Move the ruler to keep it balanced.

What You Saw

What part of the ruler rested on the spine each time?
Write your observations on your record sheet.

Think About It

Why did you have to keep moving the ruler to keep it balanced?
Write your answer on your record sheet.

Name _____

Balance

A Balancing Act

Let's Find Out

How can you keep a ruler balanced as it rests on a point?

What You Saw

Describe what part of the ruler rested on the spine.

When the ruler had no pennies:
_____

When a penny was on one end:
_____

When a penny was on both ends:
_____

When two pennies were on one end and one penny was on the other:
_____

Think About It

Why did you have to keep moving the ruler to keep it balanced?
_____
_____

Try This! Set up the experiment again. Put two pennies on one end of the ruler and one penny on the other. Instead of moving the ruler, move the two pennies instead. See where they should be to keep the ruler balanced.

Name _____

Gravity

Race to the Ground

Let's Find Out

How does gravity affect the speed of falling objects that have similar shapes?

What You'll Need

- short pencil and long pencil
- two soft balls of different sizes
- die and wooden cube
- stepstool

What to Do

Work with a partner.
1. Hold the short pencil in one hand and the long pencil in the other.
2. Stretch out your arms in front of you. Keep the pencils the same distance from the ground.
3. Drop the pencils at the same time. Have your partner watch to see which one lands first.
4. Repeat the experiment, but this time stand on a stepstool and drop the pencils.
5. Repeat steps 1–4 with the two balls.
6. Repeat steps 1–4 with the die and the wooden cube.

What You Saw

How did the objects fall? Did the two objects with similar shapes fall at different rates or at the same rate? Did dropping the objects from a higher point affect the results? Write your observations on your record sheet.

Think About It

What can you say about the speed of falling objects that have similar shapes? Write your answer on your record sheet.

© Milliken Publishing Company   MP3501

Name _____

Gravity

Race to the Ground

Let's Find Out

How does gravity affect the speed of falling objects that have similar shapes?

What You Saw

Fill in the chart to show what you observed.

Objects	First Try— Which landed first?	Second Try— Which landed first?
Short pencil, long pencil		
Two different-sized balls		
Die, wooden cube		

Did the two objects fall at different rates or at the same rate each time?

_____

Did dropping the objects from a higher point affect the results? _____

Think About It

What can you say about the speed of falling objects that have similar shapes?

_____

_____

Did You Know?
The Italian scientist Galileo lived 400 years ago. He believed that objects fell at the same speed. It is said that he dropped two cannon balls of different weights from the top of a tall tower to prove his point.

© Milliken Publishing Company 17 MP3501

Name _____

Heat

Keeping Heat In

Let's Find Out

Which is better at keeping heat from escaping—tightly wrapped layers of cloth or loosely wrapped layers of cloth?

What You'll Need

- 2 identical jars with lids
- 2 pieces of lightweight cotton cloth (long enough to be wrapped four times around the jar)
- 2 thermometers
- teapot
- hot water
- masking tape

What to Do

1. Fill a teapot with hot water from the tap.
2. Wrap one piece of cloth tightly four times around one jar. Tape it in place. Loosely wrap the other piece four times around the other jar.
3. Pour equal amounts of hot water in each jar. Measure the temperature of the water in each jar.
4. Screw on the lids of the jars.
5. Take the temperature again after half an hour and an hour.

What You Saw

Record the temperatures on your record sheet.
Which lost more heat—the jar that was tightly wrapped or the jar that was loosely wrapped? Write your answer on your record sheet.

Think About It

When you wrapped the cloth loosely around the jar, something got "trapped" between the layers. What do you think that was?
How does wearing extra layers of clothing in winter help people stay warm? Write your answers on your record sheet.

© Milliken Publishing Company

Name _____

Heat

Keeping Heat In

Let's Find Out

Which is better at keeping heat from escaping—tightly wrapped layers of cloth or loosely wrapped layers of cloth?

What You Saw

Temperature of Water

	At first	After half an hour	After an hour
Tightly wrapped jar			
Loosely wrapped jar			

Which lost more heat—the jar that was tightly wrapped or the jar that was loosely wrapped?

_____

Think About It

When you wrapped the cloth loosely around the jar, something got "trapped" between the layers. What do you think that was?

_____

How does wearing extra layers of clothing in winter help people stay warm?

_____

_____

Try This!
Get a wool sock or some wool cloth. Put it in a jar of water. Use a pencil to push the wool down. Do you see bubbles of air? Why is wool used for making winter clothes?

© Milliken Publishing Company — MP3501

Name _____

Air pressure

Pushing Power

Let's Find Out

Will two objects move apart or move together when air is blown between them?

What You'll Need

- 2 balloons
- tape
- 2 half-sheets of paper
- 2 12-inch pieces of string

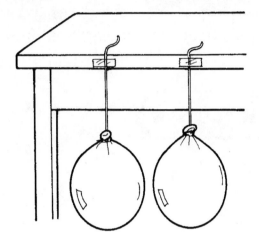

What to Do

Activity 1
1. Blow up two balloons. Tie a string to each one.
2. Tape the strings to the edge of a table so that the balloons hang two inches apart.
3. Blow air between the balloons.

Activity 2
1. Hold a sheet of paper in each hand. Bring them together in front of your mouth, but don't let them touch.
2. Blow air between the two sheets.

What You Saw

What happened when you blew air between the balloons and between the sheets of paper? Write your answers on your record sheet.

Think About It

Moving air has less pushing power than still air has. That means it does not press on objects as much. What happens when air is blown between two objects, and why? Write your answer on your record sheet.

© Milliken Publishing Company

Name _____

Air pressure

Pushing Power

Let's Find Out

Will two objects move apart or move together when air is blown between them?

What You Saw

What happened when you blew air between the two balloons?

_____

What happened when you blew air between the two sheets of paper?

_____

Think About It

Moving air has less pushing power than still air has. That means it does not press on objects as much. What happens when air is blown between two objects, and why?

_____
_____
_____
_____

Try This!
Get a 2" x 8 1/2" paper strip. Hold the strip in front of your mouth. Blow over the top of the strip. See what the strip does. Can you explain what happens?

© Milliken Publishing Company   21   MP3501

Name _____   Path of air

Hidden Target

Let's Find Out

Can you move a paper strip that is hidden behind a bottle?

What You'll Need

- clear, plastic 2-liter soda bottle with its label taken off
- 1" x 4" strip of notepaper
- tape

What to Do

1. Fold up half an inch from one end of the strip of paper.
2. Tape the paper onto a table so that it stands up.
3. Place the soda bottle between you and the strip.
4. Blow directly at the bottle in front of you. Look through the bottle as you blow, and watch what happens to the strip.

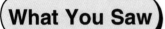

What You Saw

What happened to the paper strip when you blew on the bottle?

_____

Think About It

What happened to your breath of air once it left your mouth and struck the bottle?

_____

_____

Try This! Do the activity again, but move the strip farther behind the bottle. How far away can you place the paper strip and still make it move by blowing on the bottle?

© Milliken Publishing Company

Name _____   Water movement

Flower Power

Let's Find Out

Why do the petals of a paper flower "open up" in water?

What You'll Need

- six-inch square of photocopy paper
- pencil
- scissors
- bowl of water

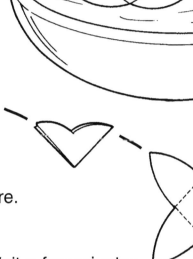

What to Do

1. Fold the paper into fourths. Draw a curve *V* as shown in the picture.
2. Cut along the lines. Unfold the paper.
3. Fold each "petal" up.
4. Place the flower in a bowl of water. Wait a few minutes.

What You Saw

What happened to the flower's petals?

_____

_____

Think About It

What do you think pushed the petals open?

_____

_____

Try This! Make three more flowers, but use photocopy paper, construction paper, and cardstock. Set each flower in a bowl of water. See which flower opens up first.

Name _____   Rocks—hardness

Hardness Test

Let's Find Out

How can you compare the hardness of different types of rocks?

What You'll Need

- 3 different rocks
- penny
- nail
- damp cloth
- magnifying glass

What to Do

1. Scratch each rock with a fingernail.
   Each time after you scratch, wipe the rock with a damp cloth.
   Use a magnifying glass to check if a mark has been left on each rock.
2. Repeat step 1, but use a penny to scratch the rocks.
3. Repeat step 1, but use a nail to scratch the rocks.
4. Scratch each rock with the other rocks. Wipe the rock with a damp cloth after each test. Use the magnifying glass to see if a mark has been left.

What You Saw

Draw your rocks on your record sheet.
Write which items were able to scratch each rock.

Think About It

Did the hardness of your rocks vary? How could you tell?
Which rock was the hardest? Which one was the softest?
Write your answers on your record sheet.

© Milliken Publishing Company

Name _____   Rocks—hardness

Hardness Test

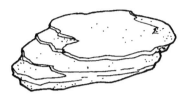

Let's Find Out

How can you compare the hardness of different types of rocks?

What You Saw

Draw your rocks in the boxes.

Rock 1   Rock 2   Rock 3

Did the rocks get scratched? Write **yes** or **no** in the chart.

	Fingernail	Penny	Nail	Rocks	
Rock 1				Rock 2	Rock 3
Rock 2				Rock 1	Rock 3
Rock 3				Rock 1	Rock 2

Think About It

Did the hardness of your rocks vary? How could you tell? _____

_____

Which rock was the hardest? Which one was the softest? _____

_____

Name _____   Rocks and minerals

Bubbling Rocks

Let's Find Out

Which rocks contain the mineral calcite?

What You'll Need

- seashell
- chalk
- limestone
- 2 different rocks (such as granite or slate)
- clear, plastic cup
- plastic spoon
- paper towels
- magnifying glass
- vinegar

What to Do

Calcite dissolves in acid. Shells contain calcite. See what happens when a shell is put in vinegar, a weak acid. Use the result to test for calcite in rocks.

1. Fill a plastic cup half full with vinegar.
2. Put the shell in the vinegar. Look at the shell with the magnifying glass. See if any bubbles form.
3. Use the spoon to take the shell out of the cup. Leave the shell on a paper towel to dry.
4. Repeat steps 2 and 3 with the chalk, limestone, and other rocks.

What You Saw

Write what happened on your record sheet.

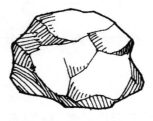

Think About It

Which of your samples contain calcite? Write your answer on your record sheet.

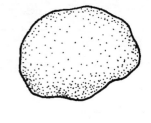

© Milliken Publishing Company

Name _____

Rocks and minerals

Bubbling Rocks

Let's Find Out

Which rocks contain the mineral calcite?

What You Saw

Write what happened when you put each sample in the vinegar.

	What happened in the vinegar?
Shell	
Chalk	
Limestone	
Rock 1	
Rock 2	

Think About It

Which of your samples contain calcite?

_____
_____
_____

Did You Know?
Chalk is a type of limestone. it is formed from the shells of tiny sea animals. In certain kinds of limestone, you can see the shells that make up the rock.

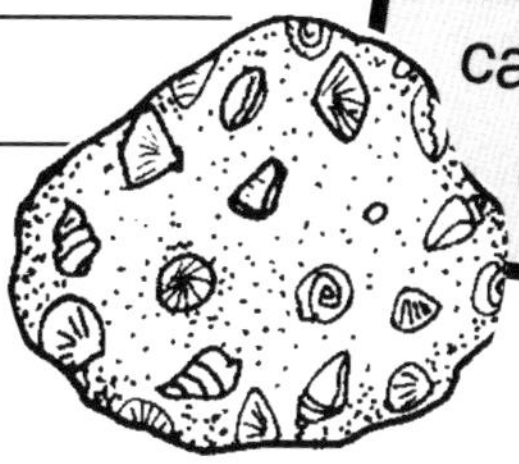

© Milliken Publishing Company   27   MP3501

Name _____

Thermometers

Temperature Ups and Downs

Let's Find Out

How does a thermometer work?

What You'll Need

- outdoor thermometer
- paper towel
- water

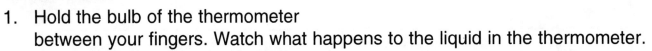

What to Do

1. Hold the bulb of the thermometer between your fingers. Watch what happens to the liquid in the thermometer.
2. Wet a paper towel with cold water. Fold the paper towel into a small square. Hold the bulb between your fingers and press the paper towel gently against it. Watch what happens.

What You Saw

What happened to the thermometer when you first held it between your fingers?

_____

What happened when you pressed the cold paper towel against the thermometer?

_____

Think About It

Why did the liquid in the thermometer go up and down?

_____

_____

© Milliken Publishing Company

MP3501

Name _____

Weathering

Rocky Breakup

Why do rocks high up in the mountains sometimes break apart?

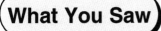

- plastic container with lid (such as a cottage cheese container)
- water

What to Do

1. Fill the container to the top with water. Snap on the lid.
2. Place the container in the freezer for 24 hours.
3. Take the container out of the freezer and examine it.

What You Saw

What did the container look like when you took it out of the freezer?

_____

_____

Did You Know? When water freezes, it expands with great force. It can exert a pressure of more than 4,600 pounds per square inch. That's the kind of pressure you'd get from an elephant standing on a postage stamp!

Think About It

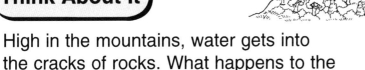

High in the mountains, water gets into the cracks of rocks. What happens to the rocks when the water freezes?

_____

_____

© Milliken Publishing Company
MP3501

Name _____   Erosion—wind

Blown Away

Let's Find Out

How do plants affect the way soil is blown away by wind?

What You'll Need

- two foil pie plates
- trowel or toy shovel
- dry dirt
- stiff leaves (or craft sticks)
- straw

What to Do

1. Go outdoors onto a grassy area.
2. Fill each pie pan with dirt. Make two mounds. Insert leaves into one mound.
3. Blow through the straw at the mound that has no leaves. Watch what happens to the dirt.
4. Blow through the straw at the second mound. Watch what happens to the dirt.

What You Saw

What happened to the dirt in the mound with no leaves?
What happened to the dirt in the mound with leaves?
Which mound lost more dirt?
Write your answers on your record sheet.

Think About It

Which would lose more soil by wind—a hill with no plants or a hill covered by grasses, bushes, or trees? Why?
Write your answers on your record sheet.

© Milliken Publishing Company   MP3501

Name _____   Erosion—wind

Blown Away

Let's Find Out

How do plants affect the way soil is blown away by wind?

What You Saw

What happened to the dirt in the mound with no leaves?

_____

What happened to the dirt in the mound with leaves?

_____

Which mound lost more dirt? _____

Think About It

Which would lose more soil by wind—a hill with no plants or a hill covered by grasses, bushes, or trees? Why?

_____
_____
_____

Try This!
Repeat the experiment to test how plants affect the way soil is washed away by rain. Instead of blowing at the mounds, use a watering can to sprinkle water over them. See which mound loses more soil.

Name _____

Sand dunes

Hills of Sand

Let's Find Out

How are sand dunes formed?

What You'll Need

- shallow baking pan
- straw
- sand
- paper cup
- newspaper

What to Do

1. Work on the floor or at a table. Protect your work area by covering it with newspaper.
2. Put the pan on the newspaper. Use the cup to fill the bottom of the pan with sand.
3. Blow at the sand through a straw. Notice what happens to the sand.
4. Continue blowing until you have made several "hills."

What You Saw

What happened to the sand when you first blew at it with a straw?
What happened as you kept on blowing? Write your answers on your record sheet.

Think About It

Dunes are hills of sand. They are found in sandy deserts and other places where there is a lot of sand.
How are sand dunes formed? What do you think happens to a dune when the direction of the wind changes?
Write your answers on your record sheet.

© Milliken Publishing Company

MP3501

Name _____

Sand dunes

Hills of Sand

Let's Find Out

How are sand dunes formed?

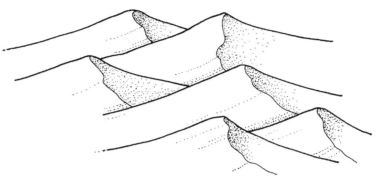

What You Saw

What happened to the sand when you first blew at it with a straw?

_____

What happened as you kept on blowing?

_____
_____

Think About It

Dunes are hills of sand. They are found in sandy deserts and other places where there is a lot of sand. How are sand dunes formed?

_____
_____
_____

What do you think happens to a dune when the direction of the wind changes?

_____
_____

Did You Know?
Africa's Sahara Desert is the world's largest desert. It has strong winds and vast "seas of sand." The winds are so powerful that dust from the Sahara has traveled as far away as Germany and Great Britain!

© Milliken Publishing Company 33 MP3501

Name _____   Erosion—glaciers

Rivers of Ice

Let's Find Out

How do glaciers affect the earth's surface as they move?

What You'll Need

- ice cube
- small plate of sand
- wax paper

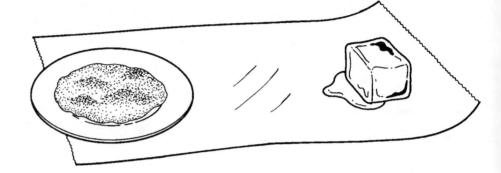

What to Do

1. Rub your finger over the ice cube so that it begins to melt. Rub the ice cube against the wax paper. Check the surface of the wax paper.
2. Dip the ice cube into the sand. Rub the sandy side of the ice cube against the wax paper. Check the surface of the wax paper.

What You Saw

What happened to the wax paper when only the ice rubbed against it?

_____

What happened to the wax paper when the ice and sand rubbed against it?

_____

Think About It

Glaciers are huge rivers of ice. A glacier picks up pieces of rock as it moves. What happens to the earth's surface as the glacier moves?

_____

Did You Know? Glaciers are found in polar regions and in mountains. One-tenth of the earth's land is covered by glaciers and ice.

© Milliken Publishing Company   34   MP3501

Name _____

Volcanoes

Watch an Eruption

Let's Find Out

What causes volcanoes to erupt?

What You'll Need

- clear salad dressing bottle
- large plastic tub
- baking soda
- 1/4 cup of vinegar
- warm water
- funnel
- teaspoon
- red food coloring

What to Do

1. Put the bottle in the tub. Fill it halfway with warm water. Add a drop of food coloring. Shake gently to mix.
2. Put a funnel in the bottle's opening. Add 3 heaping teaspoons of baking soda.
3. Pour the vinegar into the bottle.

What You Saw

What happened when you added the vinegar?

_____

Think About It

Mixing baking soda and vinegar created a gas called carbon dioxide. The gas expanded (it took up more space). How did this affect the water in the bottle?

_____

Hot, liquid rock called magma lies deep beneath a volcano. Magma contains carbon dioxide and other gases. What happens when the gases expand?

_____

© Milliken Publishing Company   MP3501

Name _____

Earthquakes

Earthquake Shake

Let's Find Out

How does the distance from an earthquake's source of shaking affect how much damage a place gets?

What You'll Need

- box (about 12" x 18" on its bottom side)
- 6 small rectangular blocks

What to Do

1. Place the box bottom-side up on the floor.
2. Stack three blocks to make a "building."
   Stack the other three blocks to make a second building.
3. Put one building about three inches from the edge of the box.
   Put the other building about three inches from the opposite edge of the box.
4. Place your fingers near one building. Start tapping on the box.

What You Saw

What happened to the buildings when you tapped the box?

_____

Think About It

Which building got more "damage"? Why do you think this happened?

_____

How does the distance from an earthquake's source of shaking affect how much damage a place gets?

_____

Name _____

Rainbows

Catch a Rainbow

Let's Find Out

What causes a rainbow to appear?

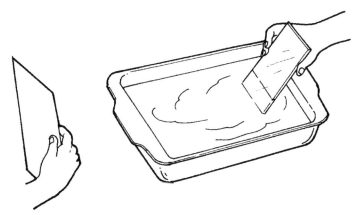

What You'll Need

- baking pan
- small, flat mirror
- sheet of white paper
- water

What to Do

Work with a partner. Do this activity either outdoors on a sunny day or indoors near a sunny window.
1. Put the pan on the ground or on a table. Fill the pan with water.
2. One person holds the mirror in the water. The mirror should be held at a slant so that the sun shines on the part of it that is in the water.
3. The other person holds the sheet of paper. It should be held so that the reflection from the sunlight hitting the mirror can be seen on the paper. The paper may need to be moved away from the pan a bit.

What You Saw

What did you see on the paper when it caught the reflection?

Think About It

Sunlight may look invisible, but it is really made up of different colors. When sunlight passes through water at a slant, you can sometimes see those colors. What colors are they?

© Milliken Publishing Company

MP3501

Name _____

Clouds

Cloud in a Jar

Let's Find Out

How are clouds formed?

What You'll Need

- clear jar
- warm water
- small container that will sit on top of the jar
- ice cubes

What to Do

1. Place several ice cubes in the small container.
2. Pour a small amount of warm water into the jar.
   Put in just enough so that the water level is about one inch from the bottom.
3. Set the container of ice cubes on the jar.
   Watch for a few minutes to see what happens inside the jar.
4. After 10 minutes, lift the container of ice cubes and look at its underside.
5. Put the container back on the jar. Check the container after half an hour.
   Look at the underside of the container again.

What You Saw

Write your observations on your record sheet.

Think About It

Why do you think the jar got "cloudy"?
Where did the moisture on the underside of the container come from?
Write your answers on your record sheet.

© Milliken Publishing Company

MP3501

Name _____   Clouds

Cloud in a Jar

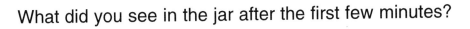

Let's Find Out

How are clouds formed?

What You Saw

What did you see in the jar after the first few minutes?

_____

What did the underside of the container look like when you first picked it up?

_____

What did the underside of the container look like when you picked it up the second time?

_____

_____

Think About It

Why do you think the jar got "cloudy"?

_____

_____

Where did the moisture on the underside of the container come from?

_____

_____

Did You Know?
Aristotle wrote the first book about weather. He lived in Greece more than 2,000 years ago. Aristotle thought that clouds were formed when fire and water mixed!

© Milliken Publishing Company   MP3501

Name _____

Sun and shadows

A Shadow Clock

Let's Find Out

How does the sun's changing position in the sky affect shadows?

What You'll Need

- paper plate
- 2 pencils

What to Do

Work with a partner.
1. Go outside on a sunny morning.
   Find a grassy spot or dirt-covered area that gets sunlight all day.
2. Poke a pencil through the center of the plate. Put the pencil into the ground.
3. Look at the pencil's shadow that falls on the plate. With another pencil, draw a line along the shadow. Write the time on the line.
4. Go out every hour and repeat the activity until you have drawn four lines.

What You Saw

When was the pencil's shadow the shortest? When was it the longest?
How did the shadows change during the day?
Write your answers on your record sheet.

Think About It

How do you think the pencil's shadow would have looked one hour earlier and one hour later? What caused the shadows to change? Write your answers on your record sheet.

© Milliken Publishing Company

MP3501

Name _____

Sun and shadows

A Shadow Clock

Let's Find Out

How does the sun's changing position in the sky affect shadows?

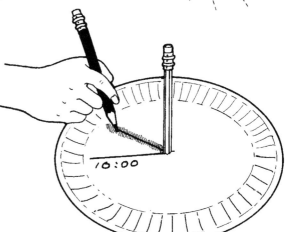

What You Saw

When was the pencil's shadow the shortest?

_____

When was the pencil's shadow the longest?

_____

How did the shadows change during the day?

_____

Think About It

How do you think the pencil's shadow would have looked one hour earlier and one hour later?

_____
_____

What caused the shadows to change?

_____

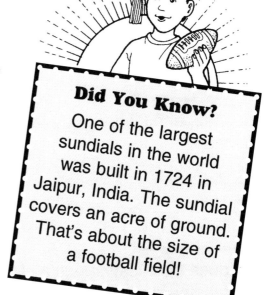

Did You Know? One of the largest sundials in the world was built in 1724 in Jaipur, India. The sundial covers an acre of ground. That's about the size of a football field!

© Milliken Publishing Company

Name _____

Fresh water and salt water—density

Floating Egg Trick

Let's Find Out

Is it easier to float in a salty ocean or in a freshwater lake?

What You'll Need

- 2 tall, clear glasses
- raw egg
- salt
- spoon
- water

What to Do

1. Fill each glass half full of water.
2. Add three spoonfuls of salt to one glass. Stir to dissolve the salt. Wait an hour until the salt water doesn't look so cloudy anymore.
3. Using the spoon, lower the egg into the fresh water. Watch what happens.
4. Use the spoon to lift the egg out of the fresh water. Lower the egg into the salt water. Watch what happens.
5. Keep the egg in the salt water. Slowly pour fresh water into the salt water by dribbling it along the inside of the glass. This will help keep the liquids from mixing. See if you can make the egg float in the middle of the glass!

What You Saw

What happened when you placed the egg in fresh water?
What happened when you placed the egg in salt water?
Write your answers on your record sheet.

Think About It

How does the "floating egg trick" work?
Is salt water more dense (heavier) or less dense (lighter) than fresh water?
Is it easier for a person to float in a salty ocean or a freshwater lake? Why?
Write your answers on your record sheet.

© Milliken Publishing Company

Name _____

Fresh water and salt water—density

Floating Egg Trick

Let's Find Out

Is it easier to float in a salty ocean or in a freshwater lake?

What You Saw

What happened when you placed the egg in fresh water?

_____

What happened when you placed the egg in salt water?

_____

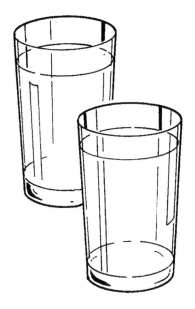

Think About It

How does the "floating egg trick" work?

_____

_____

Is salt water more dense (heavier) or less dense (lighter) than fresh water?

_____

Is it easier for a person to float in a salty ocean or in a freshwater lake? Why?

_____

_____

Did You Know?
People have been getting salt from the sea for thousands of years. In ancient times, salt was as valuable as gold. In China, people even made coins out of salt and used them for money!

© Milliken Publishing Company

Name _____   Plants—root growth

Watching Roots

Let's Find Out

Do plant roots always grow in the same direction, even when the seeds are in different positions?

What You'll Need

- 6 beans
- small cup
- jar
- paper towels
- water

What to Do

1. Soak the beans overnight in a cup of water.
2. Fold a paper towel in half and wet it. Line the inside of the jar with it.
3. Crumple up a paper towel and place it in the jar to help hold the wet paper towel in place. Moisten the scrunched-up paper towel with water.
4. Put the beans between the wet paper towel and the jar. Place them about two inches from the top of the jar. Arrange the beans in different positions: pointed up, pointed down, sideways, and at an angle.
5. Observe the beans for a week. Keep the paper towels in the jar damp but not dripping wet.

What You Saw

Write your observations on your record sheet.

Think About It

In which direction do plant roots grow? Do plant roots always grow the same way? Write your answers on your record sheet.

© Milliken Publishing Company

Name _____   Plants—root growth

Watching Roots

Let's Find Out

Do plant roots always grow in the same direction, even when the seeds are in different positions?

What You Saw

Describe what the beans looked like every other day.

Day 1	
Day 3	
Day 5	
Day 7	

Which way did the plant roots grow? _____

Think About It

In which direction do plant roots grow?

_____

Do plant roots grow the same way even when the positions of the beans are different?

_____

Try This! Place the jar of beans on its side. Watch what happens for several more days. See if the roots continue to grow in the same direction.

© Milliken Publishing Company   45   MP3501

Name _____

Seed growth—
temperature

Seeds and Warmth

Let's Find Out

Do seeds need warmth to grow?

What You'll Need

- 8 pinto beans
- small cup
- 2 self-sealing sandwich bags
- paper towels
- water

What to Do

1. Soak the beans overnight in a cup.
2. Fold each paper towel in half twice and wet them. Place each one in a sandwich bag.
3. Spread out four beans on each paper towel. Close the bags.
4. Put one bag in a warm spot. Put the other bag in the refrigerator.
5. Check the bags every other day for one week.

What You Saw

Write your observations on your record chart.

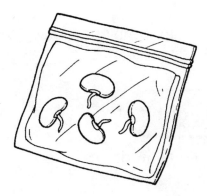

Think About It

At the end of the week, how did the beans in the warm spot compare with the beans in the refrigerator? Do seeds need warmth to grow?
Write your answers on your record sheet.

Name _____

Seed growth— temperature

Seeds and Warmth

Let's Find Out

Do seeds need warmth to grow?

What You Saw

Describe what the beans looked like every other day.

	Beans in Warm Spot	Beans in Refrigerator
Day 1		
Day 3		
Day 5		
Day 7		

Think About It

At the end of the week, how did the beans in the warm spot compare with the beans in the refrigerator?

_____

_____

Do you think most seeds need warmth to grow? _____

© Milliken Publishing Company

Name _____

Parts of a flower

Flower Detective

Let's Find Out

What are the parts of a flower?

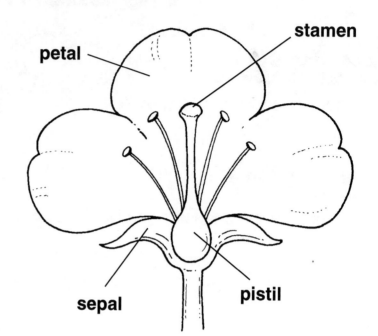

What You'll Need

- simple flower (such as a tulip, lily, poppy, buttercup, azalea, pansy, or daffodil)
- magnifying glass

What to Do

1. Look at your flower.
   Compare it to the diagram.
   (Some flowers may not have all the parts shown here.)
2. Count the **petals**. The colors of the petals attract insects and birds.
3. Find the **sepals**. They look like small leaves.
   The sepals protect the inner parts of the flower when it is a bud.
4. Count the **stamens**. The stamens produce powdery grains called pollen.
   Use a magnifying glass to examine the pollen.
5. Gently pull the petals apart. Find the **pistil**. This is the part where seeds develop and grow.

What You Saw

Draw your flower on your record sheet and write your observations.

Think About It

What are three ways flowers are the same? What are three ways they are different? Write your answers on your record sheet.

© Milliken Publishing Company

Name _____   Parts of a flower

Flower Detective

Let's Find Out

What are the parts of a flower?

What You Saw

Draw a picture of your flower in the box. Color it and label its parts.

How many petals are there? _____

How many sepals are there? _____

How many stamens are there? _____

(If there are too many petals, sepals, or stamens to count, just write **many**.)

Think About It

What are three ways flowers are the same? What are three ways they are different?

_____
_____
_____
_____

Did You Know?
The largest flower is the rafflesia. It can grow up to 3 feet wide. You probably wouldn't want this flower around, though, because it has a terrible smell!

© Milliken Publishing Company   MP3501

Name _____   Plants—transpiration

Losing Water

Let's Find Out

What happens to the water that a plant takes in but can't use?

What You'll Need

- potted plant
- plastic sandwich bag
- twist-tie

What to Do

1. Put the plastic bag over one leaf. Tie the bag around the stem of the leaf with the twist-tie.
2. Place the plant in a sunny spot for two to three hours.

What You Saw

What did you see inside the bag after a few hours? _____

_____

Think About It

A plant takes in water through its roots. The water moves up the stem to the leaves. What happens to the water that a plant doesn't need?

_____

_____

Did You Know?
Plants lose a lot of water through the pores in their leaves. On a hot day, a corn plant may lose about 4 quarts of water. That's as much as a gallon-size jug of milk!

© Milliken Publishing Company   50   MP3501